OUR PLANET EARTH

Coral Reefs

by Karen Latchana Kenney

BLASTOFF! READERS 3

BELLWETHER MEDIA • MINNEAPOLIS, MN

 **Blastoff! Readers** are carefully developed by literacy experts to build reading stamina and move students toward fluency by combining standards-based content with developmentally appropriate text.

 Level 1 provides the most support through repetition of high-frequency words, light text, predictable sentence patterns, and strong visual support.

 Level 2 offers early readers a bit more challenge through varied sentences, increased text load, and text-supportive special features.

 Level 3 advances early-fluent readers toward fluency through increased text load, less reliance on photos, advancing concepts, longer sentences, and more complex special features.

★ **Blastoff! Universe**

Reading Level

 Grade K

 Grades 1–3

 Grade 4

3 5944 00152 7645

This edition first published in 2022 by Bellwether Media, Inc.

No part of this publication may be reproduced in whole or in part without written permission of the publisher. For information regarding permission, write to Bellwether Media, Inc., Attention: Permissions Department, 6012 Blue Circle Drive, Minnetonka, MN 55343.

Library of Congress Cataloging-in-Publication Data

LC record for Coral Reefs available at: https://lccn.loc.gov/2021045039

Text copyright © 2022 by Bellwether Media, Inc. BLASTOFF! READERS and associated logos are trademarks and/or registered trademarks of Bellwether Media, Inc.

Editor: Kieran Downs   Designer: Laura Sowers

Printed in the United States of America, North Mankato, MN.

Table of Contents

What Are Coral Reefs?	4
Plants and Animals	12
People and Coral Reefs	16
Glossary	22
To Learn More	23
Index	24

What Are Coral Reefs?

← corals

Coral reefs are like long, living rocks. They are built by corals. These small ocean animals live in groups. Their skeletons build up to make reefs.

Many other animals live by coral reefs. They are busy ocean **habitats**.

Coral reefs stretch along coasts in **tropical** areas. They form when corals attach to rocks.

Great Barrier Reef

Famous For
- Largest coral reef system in the world
- Can be seen from space

Type

barrier reef

Australia → Great Barrier Reef

Size
- 1,429 miles (2,300 kilometers) long

Coral reefs grow in shallow water that is clear and warm. Sunlight shines on the corals.

polyps

Thousands of tiny coral **polyps** form reefs. They stick to rocks and grow hard skeletons.

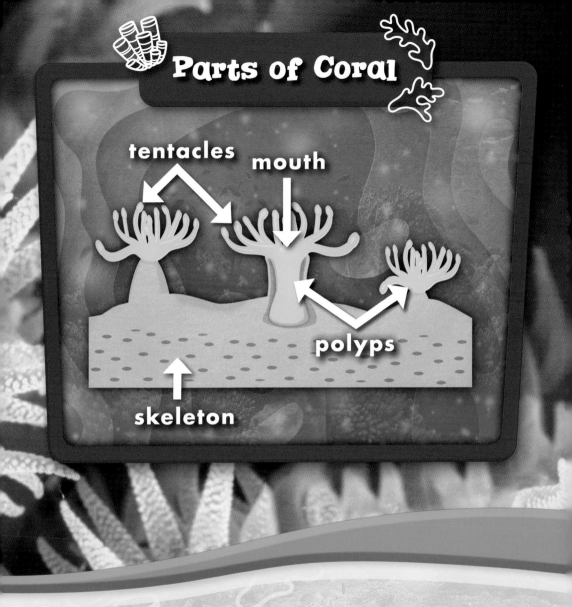

Parts of Coral

- tentacles
- mouth
- polyps
- skeleton

Polyps come in many shapes and sizes. They have soft bodies and **tentacles**. The tentacles pull food to their mouths.

fringing reef

atoll

There are many kinds of coral reefs. Three are most common.
Fringing reefs grow along shores.
Barrier reefs grow farther out.

Atolls form a ring of coral around water called a **lagoon**.

Mesoamerican Reef

Famous For

- Largest coral reef in the western part of the world
- Home to more than 500 kinds of fish
- Largest number of manatees live there

Size

- Around 621 miles (1,000 kilometers) long

Mesoamerican Reef → Central America

Type

barrier reef

Plants and Animals

clownfish

sea anemone

Many plants and animals live by coral reefs. Baby lobsters hide in sea grasses. Sea anemones protect clownfish.

Green sea turtles and stoplight parrotfish eat **algae**. Sea horses move through the corals.

algae

green sea turtle

plankton

Tiny algae live inside clear polyp bodies. They make food and **oxygen** for polyps. **Plankton** are also key parts of coral reefs.

Sharks swim through reefs to find food. Moray eels hunt in between rocks.

Coral Reef Animals

yellow-mouth moray eel

pygmy sea horse

Caribbean reef shark

stoplight parrotfish

People and Coral Reefs

Many people use coral reefs. They fish for food. They make medicine from reef plants and animals.

People also visit coral reefs to see their beauty.

fishing

pollution

Coral reefs are in danger. Divers hurt reefs by touching them. **Pollution** from land gets into reefs.

Climate change is making ocean water warmer. This makes corals release their algae. They turn white. Some reefs die.

How People Affect Coral Reefs

- Divers touch corals and damage them

- Warming water causes corals to release their algae

- Pollution from land gets into reefs

People can help protect coral reefs. They can keep beaches and oceans clean. Divers can help reefs by not touching them.

Many plants and animals need coral reefs to survive. Keep this important ocean habitat alive and growing!

Glossary

algae—plants and plantlike living things; most kinds of algae grow in water.

atolls—rings of coral that make islands in the ocean with a lagoon inside

barrier reefs—coral reefs that are separated from the shore by a deep area of water

climate change—a human-caused change in Earth's weather due to warming temperatures

fringing reefs—coral reefs that form near coasts

habitats—lands with certain types of plants, animals, and weather

lagoon—a shallow body of water blocked from the ocean by a coral reef, atoll, or island

oxygen—a gas that animals need to breathe

plankton—ocean plants or animals that drift in water; most plankton are tiny.

pollution—substances that make the earth dirty or unsafe; pollution usually comes from human actions.

polyps—small, soft ocean animals; coral polyps make coral reefs.

tentacles—long, bendable parts on some ocean animals

tropical—related to a warm place near the equator

To Learn More

AT THE LIBRARY

Esbaum, Jill. *Coral Reefs.* Washington, D.C.: National Geographic Kids, 2018.

Person, Stephen. *The Coral Reef: A Giant City Under the Sea.* New York, N.Y.: Bearport Publishing, 2020.

Shaffer, Lindsay. *Clownfish.* Minneapolis, Minn.: Bellwether Media, 2020.

ON THE WEB

FACTSURFER

Factsurfer.com gives you a safe, fun way to find more information.

1. Go to www.factsurfer.com.

2. Enter "coral reefs" into the search box and click 🔍.

3. Select your book cover to see a list of related content.

Index

activities, 16
animals, 4, 5, 6, 7, 8, 9, 11, 12, 13, 14, 15, 16, 19, 21
atolls, 10, 11
barrier reefs, 10
bodies, 9, 14
climate change, 19
coasts, 6
colors, 19
effects, 18, 19
food, 9, 14, 16
formation, 4, 6, 8
fringing reefs, 10
Great Barrier Reef, 6
habitats, 5, 21
lagoon, 11
medicine, 16
Mesoamerican Reef, 11
oxygen, 14

parts, 9
people, 16, 18, 19, 20
plants, 12, 13, 14, 16, 19, 21
pollution, 18
polyps, 8, 9, 14
rocks, 4, 6, 8, 15
sizes, 9
skeletons, 4, 8
sunlight, 7
tropical, 6
water, 7, 11, 19

The images in this book are reproduced through the courtesy of: Rich Carey, cover; arka38, p. 3; Madelein Wolfaardt, pp. 4-5; Solarisys, p. 5; JC Photo, p. 6; Anita Kainrath, pp. 6-7; Konstantin Novikov, pp. 8-9; Fabio Lamanna, p. 10 (atoll); Tanya Puntti, pp. 10-11; Tom Till / Alamy Stock Photo, p. 11; Richard Whitcombe, pp. 12-13; NaturePicsFIlms, p. 13 (turtle); Ashish_wassup6730, p. 13 (algae); Gilles Brignardello, p. 14; Choksawatdikorn, p. 14 (plankton); Richar Whitcombe, p. 15 (yellow-mouth moray eel); Dai Mar Tamarack, p. 15 (pygmy sea horse); Jesus Cobaleda, p. 15 (stoplight parrotfish); Alex Tihonovs, p. 16; Damsea, pp. 16-17; Tunatura, p. 18 (pollution); Jolanta Wijcicka, pp. 18-19; frantisekhojdysz, pp. 20-21; Kletr, p. 23.